Firefox OS for Activists

*by Eric Lee
and Jeremy Green*

Copyright 2013 Eric Lee & Jeremy Green
All rights reserved
ISBN: 1492179132
ISBN-13: 978-1492179139

Contents

Introduction 1
The problem 3
   The rise of the smartphone 3
   Smartphones are part of a closed ecosystem 4
   Web apps as an alternative 9
   The Duopoly 10
   Who's afraid of the duopoly? 15
   The phones are part of the problem 19
Fairphone 22
Open Source and Free Software 26
Meet Firefox OS 28
   Cheaper phones, easy app creation and privacy
.... 31
   The Mozilla Foundation 38
   Cheaper apps 43
   Software diversity is a good thing 45
   Ease of app creation 48
   Privacy matters 49
   What could possibly go wrong? 53
Open source alternatives to Firefox OS 57
   Sailfish 57
   Tizen 60
   Ubuntu 61
   Android Mods 65
Conclusion 70
What now? 72
Acknowledgments 74

Introduction

The internet is not what it used to be. Of course, it never is – but the changes it has gone through over the last couple of years are as profound as the creation of the World Wide Web.

Today, or in the very near future, most people will connect to the internet to read their emails, visit websites, and participate in social networks using their mobile phones or tablets.

some people will continue to access the net on the kinds of devices that already have existed for several decades, such as desktop and laptop computers. This is certainly going to be the case in offices, where desktop computers remain quite common.

But those relatively large PCs will become less important as smartphones and tablets become more popular. So the way that people use the internet will also change; it will become much more frequent, but also much more casual. Internet content is likely to be consumed through a dedicated 'app' rather than through a general-purpose web browser.

And this in turn means that the smartphone operating system becomes a gateway – or a gate-keeper – to the way that users experience the internet.

This change has implications for social change activists – trade unionists, human rights campaigners, environmentalists and others.

It will affect the way we share information, build networks, and campaign.

The purpose of this short book is to explore one of the newest operating systems for smartphones and tablets – Firefox OS – as well as other, similar, open source alternatives including Sailfish, Tizen and Ubuntu Touch.

The problem

The rise of the smartphone

At the beginning of 2012 smartphones and tablets together accounted for around 11% of all internet accesses. Just one year later that proportion had almost doubled, to around 21% of accesses.

The sci-fi dream of a tiny pocket computer, connected to all of the world's information sources, multimedia content and communications systems had been realized – along with previously unimagined kinds of usage such as social networks and microblogging.

These computers interact with us, and with the physical environment, in lots of new and interesting ways, including built-in GPS receivers, motion detectors and cameras. Oh, and they work as telephones too.

So what's not to like? For one thing, they are expensive. The average smartphone costs $375. That's a substantial drop since last year, when it cost $450, but still not cheap. They will get cheaper, of course, particularly as Chinese manufacturers with as-yet unfamiliar names crowd into the market.

Smartphones are part of a closed ecosystem

But there is another downside. Smartphones are a bit special; whereas on a computer users engage with the internet primarily through a web browser, on smartphones they are more likely to use a specialised "app" (short for application).

These apps are designed to make full use of the smartphone's capabilities (like those built in GPS receivers), but also to get around the smartphone's shortcomings. Proper web pages don't display nicely on little screens, which are often a different shape to the screen of a PC. The user has a touchscreen rather than a mouse cursor to move around the screen, and a tiny "soft" touchscreen keyboard that makes entering lots of text (or even filling in an online form) really tedious.

There's something called "responsive design", which is supposed to mean that web pages are aware of the device being used to access them and display appropriately, but it tends to produce a one-size-fits-all reformatted page that doesn't actually respond to the fact that you are on a small-screen device. It doesn't recognize that behaviour on a smartphone tends to be different. Users are less likely to scroll across pages, click on links, or download PDF attachments.

Hence the preference for dedicated apps rather than plain old web pages, even "responsive" ones.

But making an app is rather more complicated than making a web page. Some knowledge of Java programming for Android, or Objective C for Apple, is required. There are some app creation tools (App Inventor and AppsGeyser for Android, and the AppBuilder) that are designed to make it easier, but they only go so far. It's still quite difficult, and lots of experienced programmers are quite bitter about it.

In addition, there's more to getting an app on to the smartphones of your target market than just building it.

Internet users can get to any web page as long as they know the address – the URL. There are lots of points of interconnection to the internet and to the web.

But smartphone apps are distributed primarily through the app stores controlled by the main platform providers – Apple and Android (i.e. Google). It's possible to offer an Android app for download outside the Google Play Store, but the default setting for smartphones is usually to only accept apps from there. You can change the settings, but it is tricky and feels slightly scary, as if you are opening up your smartphone to dangerous 'rogue' apps and viruses. Most people won't ever do it.

Some tablets don't even give access to Google's own app store – such as the Amazon Kindle Fire, which encourages users to download only apps which Amazon itself wants them to use.

With Apple even this limited option is not really possible. Apple does run some programs and tools for distribution outside the mainstream App Store, but they are still very much under Apple's control. Being an Apple developer requires subscriptions that cost you money; not a lot of money, but still a barrier for a small activist group.

The upshot of all this is that the internet and the web are being transformed into something new, something much more commercial and structured and less open to participation by the many. Less open, too, to unions and activist groups.

This hasn't escaped the attention of some prominent internet gurus, who have deplored the way that the "generative" capabilities are being squeezed out by more closed devices and platforms. For those of historical bent, it's worth remembering that broadcast radio went through a similar transformation, starting out as a very open, community-based medium before becoming commercialized and locked down.

Even businesses, especially small and medium sized ones, struggle to make sense of the new frameworks.

Community groups, social change activists, and the labour movement will find it even harder.

There haven't been many successful apps created by or for groups like this. Although the LabourStart survey found that union members (or at least those who responded to the survey) are even more likely to access the web via smartphones and tablets than the general population, union experience with apps is at best mixed.

Only 12% of English-speaking union activists, and 8% of French-speaking activists, thought that their unions had developed any apps, though those that did were quite likely to use them, and an increasing proportion felt that the apps themselves were quite good.

A union in Sweden decided to create an app for both Apple and Android smartphones. Called "Kolleagues" it's a game that teaches workers about the need for unions. We've heard that the development of the game cost the union 25,000 Euros – none of which would have been recovered by sales, as the app is given away for free.

While the Apple iTunes store doesn't say how many copies of the app have been downloaded, the number must be fairly low as there is only one customer review. The Google Play store for Android apps claims that it has been downloaded between five and ten thousand times.

The largest public sector union in Britain, Unison, also released an app for Apple and Android devices – and Blackberry as well. When we asked union officials why they invested in Blackberry, it was so that union staffers – who were issued with Blackberry smartphones – could use the app.

The app itself is not widely used in the union and has not even been promoted to its wider membership. It does little more than give members access to what is on the union website.

The cost of making this app too was considerable – in the thousands of pounds.

Despite their development resources the big players of "clicktivism" such as Avaaz (itself popular among union members) and 38 Degrees have not created their own mobile apps, and are still relying on emails and links to their main web site; if you are one of their subscribers, then you already know that the mobile experience is not too good.

There has been talk of other groups creating their own apps, including one for 'Occupy' groups to facilitate consensus-based assembly meetings, but this does not seem to have happened. There are some apps designed specifically for activists (one sends a predefined SMS message to your lawyer when you are arrested), but despite lots of ideas the apps themselves are few and far between, and not very inspiring or useful.

Web apps as an alternative

There is an alternative to the world of dedicated, bespoke smartphone applications: web apps.

A web app is a sort of half-way house between a web page and an application. Based on standards like HTML (hyper-text markup language), JavaScript and CSS (cascading style sheets), a web app can be built using standard web creation software tools (unlike a smartphone app). Many of these are free and Open Source.

A web app works more like a traditional smartphone app in that it can be used even when the device is not connected to the internet. Browser-based apps can make some use of the smartphone's capabilities – they can tell which orientation you are using to view the web page, for example, or access other files on your device. Best of all, a web app doesn't require special client software – the browser is the client (the user-side software which accesses and presents the information in the application). So there's no need to keep lots of different versions up to date; the browser does this for you.

Like smartphone apps, web apps require that the user downloads and installs them, but this doesn't have to be done via an app store. It's possible for the user to download and install a web app from a conventional website – you may have had this experience when you've visited a website on your smartphone's browser and it tells you that there is an app available for download. If you are lucky, or proficient at controlling the settings on your smartphone OS, you may even have managed to follow through and actually install the app; then again, you may have just gone through to the "native", non-mobile website and had a mixed experience.

So why doesn't everyone love – and use – web apps?

The Duopoly

The world of internet services on mobile devices is now dominated by two great beasts – Apple and Google.

Each has created its own ecosystem which incorporates devices, operating system software, an application development framework, and an online store to sell applications and content. The two ecosystems are different from each other, of course. Apple makes most of its money from selling its fashionable, "cool" hardware, and its application and content business is an adjunct to that. Google makes its money from selling advertising on the web; hardware that runs its Android operating system is made by the major manufacturers such as Samsung and HTC, who license the operating system for free.

The two systems are run in quite different ways. Android (originally a separate company that was bought by Google in 2005) is based on Linux and is (partially) open source.

The Android apps store (now known as Google Play) has historically taken a relatively relaxed attitude towards developers, with a light touch on quality control and many applications provided for free.

On the other hand, Apple's iPhone Operating System (iOS) is proprietary, and Apple has taken a fiercely protective attitude towards the contents of its apps store – in terms of quality, and also considering the impact of particular kinds of content on its brand.

This contrast rather implies that Google is the good guy here – cheaper hardware and more hardware choice on Android, open source software, free licensing for the operating system, free applications, laid-back apps store rules and financial terms.

But it's important not to get carried away. Both Google and Apple control their ecosystems. And if it feels like you don't pay for all that stuff from Google, that's because you aren't really Google's customer. Its customer is the advertiser – you, and your attention, is what Google is selling.

To make your attention more valuable to its customers, Google collects lots of information about you – what you search for on the internet, where you do it from, what YouTube videos you watch, what content you create and post, what you email and Tweet about, who you email and message to, who is in your address book and so on.

This does not mean that Google's top management are taking a direct personal interest in you, and that they or their minions are sitting down to read your emails. They don't have to, because they have software for that.

And of course Google isn't the only one doing this. Microsoft and Apple, and others, all do the same kinds of thing. Google just does it better, with more powerful tools and relationships.

Back in 2010 Apple ran into some trouble when it became apparent that it was selling users' location data to third parties – within the terms of its end user licence agreement, of course. You do remember signing that, don't you? At the time of writing there is a similar storm about AT&T selling users' location data.

This is not in itself something sinister, though realising that you are the product rather than the customer may come as something of a shock.

These companies are collecting data about you to sell, just as your favourite newspaper and TV station does.

But they are also providing that information, on demand, to government agencies. For the most part they are doing this because they have to; it is the law, and it's part of the price of doing business. But those government agencies have no interest in selling you anything. They really are interested in you as an individual, and in who your friends and contacts are, and what you talk to each other about. For a few months in the Summer of 2013, government programs such as the US PRISM, run by the NSA, and the support they received from the giants of the Internet, were headline news. The British and US governments struggled to establish that the surveillance they carried out, without specific warrants, was within the law. Then everyone lost interest and the programs remained in place.

And that is when the Apple-Google duopoly starts to look less attractive. If you decide to move to another platform, which might be less compliant to the demands of government security agencies, you don't have a lot of choice. There are other operating systems, but their market share is negligible. Android and Apple iOS have 92% of the smartphone operating system market.

Others have struggled to make inroads. Microsoft, despite its dominance of the PC operating system market, has not managed to gain a substantial foothold so far. Nokia, until a few years ago the strongest player in the mobile handset business, backed the wrong operating system (for the best of reasons) and has been sidelined by the upstarts from outside the phone world. Blackberry, which almost invented the smartphone category for the business customer and briefly showed signs of breaking into the consumer market, now languishes around the 3% mark. The rule of two – the duopoly – seems hard to break.

The Apple world is known to be proprietary and closed. The company maintains tight control over hardware, software, and applications. Users benefit because everything works well together without the need for much technological expertise, but they pay a price in that everything is also a bit more expensive because there is no competitive marketplace. They also pay in that Apple is a bit like the Hotel California; it's hard to leave Apple and take your content to another platform, because of all that non-standard software and hardware.

It's hard to leave Google, too, though for different

reasons. You can get your stuff out of Google, but it's so hard to summon up the effort required to find alternatives and to do the actual porting. Some of the alternatives aren't as good, and they don't play as nicely with each other as the stuff that Google offers bundled into Android. In particular, they often don't synchronize properly "in the cloud", so that any changes you make on the application in your phone also appear if you look at the same application (a "to-do list", for example) from a web browser on another device. Google's own applications are the default and the path of least resistance, and most people on Android will probably stay with them most of the time.

Who's afraid of the duopoly?

The duopoly is not good for consumers, and for anyone who doesn't like too much concentration of power at any point in the mobile internet delivery chain.

There is a little bit of upside for developers, in that a smaller number of operating systems reduces the effort that they have to make in order to customize their applications. Of course, if the operating systems themselves played nicely with open web standards this wouldn't be a problem anyway. Moreover, the fact that there are only two main platforms means that independent developers don't have a lot of bargaining power. They can take the deal that's on offer, or someone else will.

Fortunately for the rest of us, some of the big players in the telecoms industry don't like the duopoly either. The network operators are well aware that the balance of power has shifted away from them and towards the operating system players – Apple and Google.

There are lots of operators but only two major OS's, so it is the latter that have the bargaining power. When it first launched the iPhone Apple was careful to enter into time-limited exclusive arrangements with a few networks. At that point it was the only smartphone that existed. Operators fell over themselves to negotiate deals with Apple. The advent of Android-based smartphones tipped the balance back somewhat.

But the centre of gravity, and the 'ownership' of the customer, was shifting. When I put any SIM card into an Android handset, I sign in once and it's my phone, linked to all my accounts. My emails arrive, I get posts from my Facebook friends, and so on. If I change phones, or change network providers, these links come with me. If I get a new network provider I can take my telephone number with me, but even if I don't the relationships with my online accounts remain intact.

And it's the same if I get a new phone. When I switch from Samsung to HTC, or to Sony, many aspects of the experience are identical (even though the big handset makers are trying to reposition themselves as online players with their own unique "cloud-based" services). The hardware itself is less relevant. The balance of power in the "mobile internet" has become a lot more like the mainstream web, in that no-one cares much about the hardware platform (in the case of the web, the PC you use) or the connectivity provider (in the case of the web, the ISP). Both are regarded as standardised and uninteresting. It's the services that you use them to access that generate the buzz.

The main difference is that the "real" web is big and messy, with no-one really in control; but in the mobile world, the power rests with the integrated platform owners, who exert far more control over what users and the other players can do.

The network operators, and the phone manufacturers, are not at all happy about this. They have spent huge amounts of money on their brand image, and they want to remain relevant and "front of mind" for the customer. So both groups are excited about the prospects of more competition in the operating system domain.

Which brings us to Firefox OS.

On the manufacturing side Firefox OS has the support of Alcatel, Huawei, ZTE, and LG Electronics. Perhaps equally important, Foxconn (the contract manufacturer used by some of the major phone brands, but most notably by Apple) is also getting behind the Firefox OS initiative and is hiring engineers to support the development.

In terms of the network operators, Mozilla announced a group of 18 major operators at the March 2013 mobile industry show in Barcelona, including América Móvil, China Unicom, Deutsche Telekom, Etisalat, Hutchison Three Group, KDDI, KT, MegaFon, Qtel, SingTel, Smart, Sprint, Telecom Italia Group, Telefónica, Telenor, TMN and VimpelCom. Together these account for a huge proportion of the world's mobile subscribers.The role of ringmaster for this group has undoubtedly been played by Telefónica, which has been far and away the most active advocate for Firefox OS.

Once again, it bears saying that these companies are aligning themselves with Firefox OS because it makes good capitalist sense to them, not out of the goodness of their hearts or because they care passionately about internet freedom. They just don't like being on the wrong side of an uncompetitive marketplace. But whatever their motives, their behaviour is helping to create an opportunity for activists with other reasons to fear the Apple-Google duopoly.

The phones are part of the problem

Activists who use mobile devices as tools for social change should be aware that, even if they might be part of the solution, the phones themselves are also part of the problem.

Environmentally, the product lifecycle of the mobile phone is something of a disaster. Making the phones needs minerals that are obtained from dubious regimes and conflict zones, and industrial processes that incorporate dangerous and toxic chemicals which can poison the workers involved in the manufacturing process and the communities in which the factories are sited.

The sheer volume of devices built and shipped is huge. This means that disposal of products that have either ceased to work, or are still working but have been rendered "obsolete" by the advance of technology, is a huge problem. Regulation in developed countries requires that the companies that make or distribute the phones (often the mobile network operators) must take them back at end of life to ensure that they are properly disposed of. Even so, some 'recyclers' are better than others. There is now a global trade in used phones involving sale by the pallet-load in Hong Kong. Broken and barely functional phones end up in developing Africa and Asia, and at the end of their life become part of a growing mountain of toxic electronic waste. If they are broken down for components and consituent minerals it is unlikely to be in a safe factory.

As with many other electronics products, the phones themselves are made under conditions that would not be tolerated in most western countries, by workers who do not have the right or ability to join decent, independent trade unions.

For example, Apple does not manufacture its products itself but uses a Taiwanese contract manufacturer called Foxconn, mentioned earlier as one of the backers of Firefox OS. There have been many negative reports about Foxconn in the media, some driven by a rash of suicides at its factories in 2010. Allegations include long working hours, illegal use of child labour, discrimination against mainland Chinese, and low pay. Claims that the company has subsequently cleaned up its act have been undermined by conflicting reports that the "clean-ups" have been announced rather than implemented, notably one by the Washington-based Economic Policy Institute delightfully entitled "Still Polishing Apple".

But in this respect at least, there is nothing particularly special about Apple.

Samsung, one of the main manufacturers onside with the Android operating system, has been repeatedly criticized for exposing workers to deadly chemicals and carcinogens in its Korean factories, and for illegal labour practices in its factories in China.

It's not hard to find similar stories about other handset makers, though some are clearly better than others and make more effort to audit both their own factories and those of their suppliers. In almost all cases this is done as part of a "corporate social responsibility" program, aimed primarily at defending the suppliers' brands.

It is hard to find manufacturers using suppliers where workers are able to organize in proper unions. For more detail on this check out the Good Electronics website. One company has appeared recently claiming to manufacture "ethical" mobile phones. But like everything else, its claims must be reviewed critically.

Fairphone

Fairphone is a Dutch initiative to create an alternative to the decidedly "unfair" phones that are being made and sold today. It began life as a campaign and transformed itself into a company to actually make phones in a better way.

Its phone, which can already be pre-ordered online, is in many ways an improvement upon the mass-manufactured phones most of us carry around today.

Those phones are usually made with little or no concern for the environment or the well-being of the workers who make them.

Fairphone, on the other hand, aims to use "fair and confict-free resources", is committed to environmentally-friendly solutions to the problem of e-waste, and has given the phone an "open design". It promotes repairabiilty, and replacement components for almost everything are intended to be available to enable this (along with instruction manuals).

All good, but when it comes to who actually makes the phone, we run into some problems.

Originally, it seems, Fairphone aimed to find a

manufacturing partner in Europe, but gave up and selected contract manufacturer A'Hong, which has factories in Shenzen and Chongqing.

It says "Fairphone intends to manufacture in China because ... we feel our model can make a difference in improving working conditions and environmental impacts in China".

Fairphone says that in China it is committed to "creating a fund to improve worker's wages and working conditions and open discussions between workers and their employers". It also promises to publish an audit report of working conditions in its partner's factory.

It has partnered with "an independent, third-party social assessment organization to perform an assessment". The partner, TAOS Network, comes from the world of 'corporate social responsibility', and its union credentials are to say the least thin.

In our discussions with the company we found a good understanding of the limits of third-party monitoring and a willingness to work towards something better for workers. It is placing its hopes on its ability to strike a new kind of supply contract, with longer terms relationships and less emphasis on lowest costs; and it hopes that some of the benefits of this can be shared with the workforce.

We understand that it is negotiating from a position that is not strong, as a newly formed company with no track record or sales figures, and with a production run of 25,000 phones that amounts to three weeks' manufacturing.

To be fair to FairPhone, proper European manufacturing is no longer really an option, because the entire supply chain has moved to Asia, and 'making' a phone in Europe would just mean final assemby from Asian-made components.

But why choose a low-wage country that also happens to be completely union-free? Asia is full of countries that have low-wage workforces, but where there are unions that at least try to organise and represent those workers.

China is surrounded by such countries, any one of which (except North Korea) has a better record on workers' rights. Workers don't need "independent third-party social assessment organizations" and they don't need "open discussions" with their bosses. They need proper representation through free trade unions, not a company-run staff association and fund.

Fairphone admits that it didn't look at all the possible manufacturing options; as a small company with limited resources that's not so surprising, though it is still disappointing.

The people behind Fairphone are clearly well-intentioned and want to make the world a better place. The company website says "We support and actively promote the ILO conventions enshrining freedom of association and collective bargaining and we feel that in absence of this little change will happen in the mines or the factory floor."

But by opting for non-union manufacture in China, and trying to placate critics with sops like "social assessment" and "open discussion", it is pushing the serious issues into the future.

We hope that Fairphone will deepen its commitment freedom of association so as to ensure that all of its phones are made in factories where workers are free to join proper, independent trade unions, even if it can't deliver on this right now. A truly fair FairPhone would carry the one label that really mattered: a union label.

Open Source and Free Software

Firefox OS is part of a broad movement in the technology world called Open Source. It's based on the radical notion that creating a piece of software, no matter how wonderful it is, should not give you rights of ownership or control over that software. To count as truly Open Source, the software developers should make the source code available so that others can see how it works and make their own modifications, and they should make the software itself available under licensing terms that explicitly allows others to do this.

The most famous piece of Open Source software is the Linux operating system, which was initially created by a man called Linus Torvalds but has since been built on by a huge community of developers, often working for free, to create a family of different variants. Several of the Open Source mobile operating systems described in this book, including Firefox OS but also Google's Android OS, are based on Linux. Arguments as to how open they really are tend to focus on how much of the code is available and under what sorts of licence. Ironically the Apple operating system, which is a long, long way from open in its philosophy, is based on a computer programming language called

Unix – as is Linux.

As with any movement that is about principle, there are differences and disagreements. The 'Free Software' wing tends to emphasize the social and ethical aspects, while the "Open Source" wing emphasizes the practical and economic benefits of making software in this way, through community development.

Should you care about any of this, which sometimes sounds like a peculiar theological dispute between two tribes of geeks? Yes, you should – if it bothers you that the software you use (but probably don't really own, under the terms of that licence you agreed to the first time you installed it) will do all sorts of stuff that you don't know about, don't understand, but probably wouldn't want if you did. It might be providing information about your internet usage to someone who wants to keep tabs on you and your friends, or it might be collecting your data so that someone can market their products to you and your friends.

And there's another reason why you should support this kind of software; because it is living proof that it isn't only capitalism and the profit motive that can provide the incentives needed to make and improve great, technologically sophisticated products.

Meet Firefox OS

Firefox OS is a brand-new open source mobile operating system which will run on smartphones and tablets. It is based on Linux (as is Android) but what makes it special is where it comes from.

Firefox will be familiar to many of you as a very popular web browser which was, until fairly recently, the main competition for Microsoft in the field of web browsers.

Firefox OS is special in that it essentially transforms the whole experience of using a mobile phone or tablet into something very similar to using a web browser. It sees the world the way a web browser does.

The force behind Firefox OS is the non-profit Mozilla Foundation. As the slogan of the foundation puts it, "We're out to make a difference, not a profit."

This is significant because the other players in the field of mobile operating systems such as Apple, Google, Blackberry and Microsoft are all pretty much in it to make a profit.

That doesn't mean that Firefox OS isn't going to be commercially viable. After all, someone is going to make a profit if it succeeds. That's why it's important to also be aware of who the early backers of Firefox OS are and what they stand to gain.

There are basically two groups at work here.

First, there are the telephone manufacturers, which are sometimes unknown to most people, but they're the people who actually make the phones that we use.

Second, the carriers – global companies like Telefónica and many others – have begun to show a real interest in Firefox OS. Their reasons for doing so are explained above – basically, they have everything to gain from an end to the Apple/Google duopoly.

In addition to the phone manufacturers and carriers, it is likely that other players will be involved in ensuring the success of this new, alternative operating system.

Among those might be progressive governments in Latin America and elsewhere who have long shown an interest in open source technology.

The decision by Telefónica to embrace Firefox OS is significant in part because it is such an important carrier in Latin America and it would not be surprising to see national and local governments taking an interest in smartphones and tablets that are considerably less expensive than what Google and

Apple have on offer.

The first phones to come out using Firefox OS are low-end, inexpensive models aimed at people who've never purchased a smartphone before.

It's not an accident that Telefónica began promoting the phones first of all in Spain, a country with a very large population of unemployed and under-employed young people who may not be able to afford high-end smartphones and tablets. This was something of an about-turn, because Firefox OS was originally planned to debut in Latin America.

Firefox OS aims to offer pretty much everything that Google and Apple offer on their phones. You can make phone calls on it, check your emails, use the built-in camera to take photos, watch videos, listen to music, check maps, and so on. Like all modern smartphones and tablets, it has a strong social component, so apps offering access to Facebook, Twitter and other social networks come on every phone.

Even the way to get apps for your Firefox OS phone will be familiar to anyone who's owned a smartphone in the past. Firefox has set up the Firefox Marketplace, a one-stop shop for apps. Unlike Apple's App Store, however, this is not the only place to get apps for a Firefox OS phone.

If it sounds like Firefox OS is a lot like Android and iOS, that's intentional. The Mozilla Foundation is keen to make a product that can compete with the very best operating systems on the market today.

So what makes Firefox OS special, and of interest to social change activists?

Cheaper phones, easy app creation and privacy

There are three things Firefox OS does that make it appealing and interesting for progressives.

First, it will help make smartphones cheaper. That will make it possible for many more working class and poor people to own them.

Second, it will make it easier and cheaper than ever before to create apps, the little computer programmes that run on smartphones. This will make it easier and cheaper for campaigning organizations such as trade unions to have quality apps.

Finally, it will help make for a more private and secure experience when using mobile technology, breaking free of the Apple/Google duopoly and all that that entails.

Let's go over this point by point.

Making smartphones cheaper

If you knew nothing at all about mobile phones, you may think that the cost of making one consists of raw material, assembled components, labour and shipping. But actually a big chunk of the cost of a mobile phone can be the operating system installed on it.

The operating system on the popular Apple iPhones and iPads is now called iOS. We don't know how much of the cost of an iPhone is this operating system because Apple doesn't license it out to anyone.

If you want your phone to run iOS and the various applications that run on this platform, then you have to buy a phone or tablet that Apple manufactures.

(To be more precise: Apple itself doesn't actually manufacture its phones and tablets. It outsources the manufacture to companies like Foxconn, in China. That's why Apple products say "designed in California" but obviously not made there.)

There is no doubt that Apple invests a considerable amount of time and effort into developing its operating system, and it therefore represents a healthy chunk of the cost of the iPhone or iPad that you buy from them.

This is also true of Blackberry phones, made by a Canadian company which used to play such an important role in the world of smartphones. One might argue that Blackberries were the first successful smartphones. If you like the Blackberry operating system, you'll need to buy a phone made by them. Only Blackberry makes phones that run their operating systems – Blackberry OS and, more recently, Blackberry 10.

As this book is being published, Blackberry's future is very much in doubt – a victim of the Apple/Google duopoly.

But the other two big players in mobile phone operating systems today, Microsoft and Google, license out their operating systems. So manufacturers can build phones or tablets using those operating systems.

At the moment, Google does not charge manufacturers to use its operating system. This keeps the cost of Android phones lower than the cost of Apple's iPhones.

So if Google's not earning money from licensing its Android operating system, and it's not actually manufacturing phones, how does it make its money?

The answer is that it does so the same way as it does on the web, the way it always did: advertising.

A year and a half ago, ZDNet reported that "while Google makes nothing from Android directly, it makes a lot of money, about $2.5 billion a year, from pushing ads to Android-enabled devices. And that $2.5 billion is expected to double over the next 12 months, so it's in Google's interest to keep Android free for handset partners."

What Microsoft charges manufacturers to license its operating system is not public information.

While Google doesn't charge phone manufacturers for Android, both Microsoft and Google charge for the patents they own. Even manufacturers of Android phones pay several dollars per phone to Microsoft because it owns patents to several features used on the phone. The remains of Nokia is now partly a patent-owning company, as is Qualcomm, which designs (but does not make) the chips that run some of the smartest phones.

As we'll see, the various "competitors" actually pay each other all the time, invest in each others' technology and so on.

And the non-profit Mozilla Foundation which stands behind Firefox OS is not immune to this either, as it is partly funded by Google.

Apple, Microsoft and Google are massively successful companies reaping vast profits from their investments in mobile phones – even though none of them started this way.

Apple was originally a computer manufacturer, Microsoft made software, and Google was a search engine.

Today they are all heavily committed to mobile phones in general, and the mobile internet in particular.

Firefox OS on the other hand has its origins in a totally different place, and is going in a different direction, and this will affect the cost of their product.

The group behind Firefox OS is the Mozilla Foundation, a non-profit organization. The purpose of the Mozilla Foundation is to promote "openness, innovation and participation on the Internet". The Foundation says that it promotes "the values of an open Internet to the broader world".

The Foundation bases its work on a short manifesto which includes ten fundamental principles:

1. The Internet is an integral part of modern

life—a key component in education, communication, collaboration, business, entertainment and society as a whole.
2. The Internet is a global public resource that must remain open and accessible.
3. The Internet should enrich the lives of individual human beings.
4. Individuals' security on the Internet is fundamental and cannot be treated as optional.
5. Individuals must have the ability to shape their own experiences on the Internet.
6. The effectiveness of the Internet as a public resource depends upon interoperability (protocols, data formats, content), innovation and decentralized participation worldwide.
7. Free and open source software promotes the development of the Internet as a public resource.
8. Transparent community-based processes promote participation, accountability, and trust.
9. Commercial involvement in the development of the Internet brings many benefits; a balance between commercial goals and public benefit is critical.
10. Magnifying the public benefit aspects of the Internet is an important goal, worthy of time, attention and commitment.

Many of these principles mesh quite nicely with what

people on the left would advocate – in particular the notion that the internet is a "global public resource".

The Mozilla Manifesto is important because regardless of what Google, Apple, Blackberry or Microsoft say they are all about, in the end they're all about making profit.

All these companies have elaborate mission statements and claim to be making the world a better place.

Google famously has as its corporate motto "Don't be evil" and has gone into some detail in its Code of Conduct on what that means.

If these companies do good in the world, and make the world a better place, it's a by-product of their constant search for profit.

This is not an accusation – it's a statement of fact.

Google, Microsoft, Apple and Blackberry are all obligated to make the maximum profit to benefit their shareholders.

Their shareholders, not the people who use their products, come first.

A non-profit foundation should, therefore, be able to distribute a cheaper product, and to target audiences who have traditionally found things like smartphones and tablets to be too expensive.

That's why the Mozilla Foundation's role is so important.

The Mozilla Foundation

We should say a word or two about the Mozilla Foundation and where it comes from.

Twenty years ago, when the World Wide Web was still in its infancy, a very smart young man named Marc Andreessen created the first web browser for Windows which he called Mosaic.

He didn't invent the World Wide Web – that was the work of Tim Berners-Lee, a British academic based at the CERN research facility in Geneva.

Andreessen and Berners-Lee both did their pioneering work, laying the foundation for the web, outside of the corporate sphere. Both were academics and their projects – the web and the browser – were not designed to make money, and were given away for free.

(So much for the notion so beloved by the right that innovation comes only from the private sector.)

Andreessen's Mosaic was renamed Netscape and for some time it was the web browser everyone used.

Microsoft was very late to the game, not grasping the significance of the Internet. It was only in 1995 that Bill Gates famously warned his company that they were lagging far behind the upstart Netscape crowd and that the Internet may well become important.

Microsoft decided that the best way to get into the game was not to create a better browser – its Internet Explorer lagged far behind in features when compared to Netscape – but instead to take advantage of its monopoly position in the world of personal computers and embed the software in its Windows operating system, making it the default browser.

What this meant over time was that Internet Explorer's market share rose and Netscape's fell. This was known in the 1990s as the "browser wars".

It seemed as if Microsoft had won and plucky little Netscape had lost when the latter made an odd decision.

In January 1998, the company released the proprietary source code for its browser to the world.

The non-profit Mozilla Foundation took it over, and a

community of open source developers continued work on the Firefox browser and Thunderbird email client which were Mozilla's flagship software products.

Firefox was a better browser than Internet Explorer and over time its market share grew.

The Mozilla Foundation was helped by court decisions in the USA and Europe which compelled Microsoft to make it easier for people to change their default browsers and over time, the market share of alternative browsers (in particular, Firefox) rose.

It's worth noting that the Mozilla Foundation gets almost all of its money from donations by Google. This means that Google is funding a potential competitor to its own Android operating system. That's not quite as weird as it sounds – this sort of thing happens all the time in the wonderful world of technology. Google benefits from its relationship with Mozilla too (for instance, it is the default search engine in Firefox), and it might be maintaining a competitor in case someone ever charged it with having monopoly power in the mobile phone OS arena.

These days, several browsers dominate the scene.

Internet Explorer is still a major player, though hardly in the dominant role it had a decade or so ago.

Google's Chrome browser became hugely popular almost overnight.

And Mozilla's Firefox remains a major force.

On mobile devices, Apple's Safari browser is hugely popular, as is the Opera browser, made by a small Norwegian company that has carved a niche for itself making innovative browsers that never really caught on on the desktop – but have done very well on small screens.

The story of how Mozilla took on Microsoft in the 1990s and fought it to a draw is an inspiring one and gives hope that just as we benefit today from a choice of browsers, in the future there'll be a greater choice of operating systems for our mobile devices as well.

The Mozilla Foundation won't actually manufacture the phones – so along the way, companies will get involved which do need to make a profit.

Nevertheless, by taking the expensive, for-profit operating system out of the equation, Firefox OS promises cheaper phones for a much larger audience around the world.

Another reason why Firefox OS phones will be cheaper is what Brendan Eich, co-founder and CTO of the Mozilla Corporation argues: Firefox OS could make a dent in the low-end market because the Android operating system is simply too bloated to run on cheap hardware.

The very first Firefox OS phones came out in 2013 and cost as little as £60.00 (under US $97).

By comparison, the new Apple iPhone 5S sells for £549, with the 64 GB model costing over £700.

While the Apple iPhone 5S is a top-of-the-line model, there's no getting around the fact that it is about ten times as expensive as the first Firefox OS phones.

To be the fair, those first Firefox OS phones got rather poor reviews. They're cheaply made and don't show off Firefox OS's real strengths because of it.

There are plans to release far better and more powerful Firefox OS phones and tablets, and these should allow head-to-head competition with Android and iOS devices.

As we'll show later when discussing Ubuntu Touch, an alternative open source operating system, the use of free software doesn't necessarily mean cheap, under-performing phones. The model Ubuntu aims to manufacture is a top-of-the-line smartphone with specs that leave Apple and Samsung in the dust.

Cheaper apps

Not only will the phones running Firefox OS be cheaper, but it's likely that the apps they run will be cheaper to buy as well.

One reason is that the Firefox OS "ecosystem" is based on open source software, on the free standards on which the World Wide Web was built.

An app for Firefox OS is essentially a website – albeit an improved one, one specifically adapted for mobile devices.

Since the web became popular more than two decades ago, millions of people have become experienced web developers – though most will have only a rudimentary knowledge of things like hypertext markup language (HTML), cascading style sheets (CSS) and JavaScript.

Nevertheless, there are far more people around who can write code for a website in HTML than will be familiar with Java, which is the language Android apps are based on, or Objective C, which is what iOS apps use.

The ease of creation of new apps for Firefox OS is something we'll come to in a moment, but suffice it to say that an app that's easier to create, and that uses well-known, open web standards, is going to be less expensive than an app written specifically for Android or iOS.

But that's only part of the reason. Apps for Firefox OS will also be cheaper because Apple's not around to grab its share.

If you write an app for an iOS device and you want to make it available to everyone who uses iPhones and iPads, you need to make it available on the Apple App Store. To do so, you must submit the app to Apple for approval. And Apple also gets 30% of the price you charge.

That's right – Apple not only makes money by selling you an iPhone or iPad, but continues to make money every time you buy an app (or for that matter, a song, or movie, or book from their iTunes shop).

If you take a popular app like WhatsApp Messenger, which costs just $0.99, Apple will be taking $0.30 leaving only $0.69 to the developer. The price is therefore inflated by nearly half in order to ensure that Apple makes money. The developer would make the same profit selling the app for only $0.69.

WhatsApp claims over 250 million active users, and if all of them bought the app, it means the company took in about $250 million dollars – $75 million of which went to Apple.

That's a staggering amount of money. No wonder Apple wants to keep control of the system – even it costs users 50% more per app as a result.

With many more people making apps using web standards, and without Apple taking its 30% share, we can expect that apps for the open source Firefox OS system will be cheaper, and in many cases, completely free of charge.

Software diversity is a good thing

Without making a fetish of competition and free markets, even socialists accept that innovation can be a very good thing.

Just as we support biodiversity, so we should support diversity of software choices

Apple and Google's near-monopoly of apps strangles innovation, is potentially a tool for censorship, and keeps the cost of apps artificially high.

To get an app distributed in Apple's closed ecosystem, you need to register yourself as an Apple developer and pay an annual fee. Then you go through a long and convoluted process of submitting your app to the

App Store and wait patiently while Apple decides whether or not your app deserves to be on iPhones and iPads.

On the face of it, this works well. There are a staggering number of apps available and Apple can rightly claim to have created a vast industry of app developers – though initially no one other than Apple was actually allowed to create apps.

Not only does Apple control the process of app submission to its App Store, it also controls the updates to apps. So if you need to fix or change something in an app you've submitted, you need to submit your change to Apple to be approved – often leading to delays.

Apple claims that it needs to review every app for quality purposes, but this has created a system in which a single American corporation has complete control over the software on every iPhone and iPad in the world.

This means that Apple's values – made in California – are enforced around the world. For example, iTunes will not sell a song that it believes contains profanity. Apps have been banned because they showed newspapers that featured some nudity – something which would be unacceptable in California but perfectly normal in Europe.

Apple has been known to ban magazines about rival operating systems, famously blocking a Danish-based online magazine about Android. Similarly, a tech education startup had to remove references to Android before it could appear in the App store.

For campaigning organisations, there is a serious risk that Apple's App Store would reject apps that might be used for illegal purposes, such as civil disobedience.

It's not only Apple's control of the App Store that's problematic. The company also determines what are the default apps on one's phone and tablet, and can prevent users from installing other apps in their place.

The fate of the exceptionally popular SwiftKey app is a good example of how Apple limits choice and stifles innovation.

Here is how SwiftKey describes itself:

"SwiftKey ... replaces your device's on-screen keyboard to help you type in an easier, faster and less frustrating way. The app is packed full of features to help you type, from the award-winning prediction engine that learns how you write to predict your next word, to SwiftKey Flow, which lets you write whole phrases without lifting a finger."

It's a great app – but you can't use it on your iPhone or iPad because Apple has decided not to allow third party keyboards.

To sum up, both Apple and Google largely control which software you can put on your phones. Some of their competitors, such as Amazon, are even more restrictive. In fact, some manufacturers competing with Amazon to create Android tablets proudly say that you can access the entire Google Play store – which Amazon does not allow.

Even though Firefox OS has an app store all its own, it will not have the strangling effect that Apple and to a lesser extent Google have on this.

Ease of app creation

At the moment trade unions and other civil society organizations that want to have apps for mobile phones find that they have to outsource the creation of those apps even if they've spent a small fortune over the last decade creating their websites.

That's because the tools used to create a website – the building blocks such as HTML, CSS and JavaScript – are not the same used by Android and iOS.

In theory, therefore, the creation of apps on Firefox OS devices should be considerably easier.

We say "in theory" because both Apple and Google claim that app creation for their devices is also quite easy.

But as we know from personal experience, it is anything but.

The most basic web app can be created in-house, though to get one that makes clever use of the all the features of a Firefox OS phone, one would need additional skills.

Privacy matters

Most of us take steps in our daily lives to protect our privacy. We may lock our doors, or draw curtains, or hide our PIN when we use a cash machine on the street.

And when we're online, we accept that many websites require passwords and are sometimes even encrypted, such as when we make a purchase using our credit cards.

But what we may not be aware of is just how important the choice of your mobile phone operating system is when it comes to privacy.

Think about it: the browser on your desktop PC may know the websites you visit, and your internet service provider (ISP) will know this as well, but your mobile phone and the carrier you use knows a lot more.

They know where you are and where you have been. They know who you have called, and who called you and how long you spoke.

Your smartphone can probably figure out whether you are walking or driving in a car based on your GPS location or the built-in accelerometer which may be measuring how many steps you've taken and how many flights of stairs you've climbed.

And so on.

That's why the choices we make about what smartphones and tablets we use, and what software is installed on them (especially the operating system) is so important.

Every company, and even most governments, claim to care about your privacy.

So Google, Apple and all the others do claim to protect your privacy when using their phones and operating systems – as much as they can, within the law.

The Mozilla Foundation and the mobile phone manufacturers and carriers they work with make similar pledges.

Mozilla is taking advantage of its position as both a non-profit and as not-Apple and not-Google to put out its stall in the "we care about your privacy" marketplace.

Just a week after the revelations of massive US government spying on the private communications not only of its own citizens but pretty much everyone else, the Mozilla Foundation took the initiative to launch the Stop Watching Us campaign.

While the Mozilla Foundation was taking the lead on campaigning to protect the privacy of innocent civilians, its rivals in the production of mobile phone operating systems, Google, Apple and Microsoft, were coming up with explanations of what they did for the US National Security Agency and why.

It turned out that not only did these companies do what they believed they were legally obligated to do, but that they were paid millions of dollars to do so.

According to a report in the Guardian, the British newspaper which intially broke the story about PRISM, "The National Security Agency paid millions of dollars to cover the costs of major internet companies involved in the Prism surveillance program after a court ruled that some of the agency's activities were unconstitutional, according to top-secret material passed to the Guardian. The technology companies, which the NSA says includes Google, Yahoo, Microsoft and Facebook, incurred the costs to meet new certification demands in the wake of the ruling from the Foreign Intelligence Surveillance (Fisa) court."

Had whistleblower Edward Snowden not passed on details of the PRISM project to the Guardian, Google, Apple and Microsoft, which make the operating systems used on the world's smartphones and tablets, would today still be collecting data and passing it on to the NSA. And getting paid for it too.

In contrast, the Mozilla Foundation decided to do something positive. It sought out partners in civil society to campaign for privacy

Among those who have signed up to back Mozilla's campaign are the American Civil Liberties Union, the Electronic Frontier Foundation, the Free Software Foundation, Greenpeace USA, the Green Party USA, MoveOn.org, Occupy Wall Street and so on.

This is not an assurance that the Mozilla Foundation will always do the right thing, because if it or its partners (the carriers) are served with a warrant demanding that data be turned over, it isn't promising to break the law.

But at the very least, it seems to be taking privacy very seriously indeed.

What could possibly go wrong?

Good technologies do not always succeed and it's worth looking at a couple of examples that prove this point.

In the early 1990s, former Apple Executive Jean-Louis Gassée headed up a company that aimed to produce a third operating system to rival the Apple Mac and Microsoft Windows products. It was called BeOS and was widely seen to be a first-rate product.

The company owners tried to sell it to Apple, which at the time was faltering, but couldn't agree a price (Apple was reported to have offered $125 million; Be Inc. wanted $200 million). The two companies went their own ways. Apple has gone on to become one of the most successful companies in history. Be Inc. disappeared within a few years.

Enthusiasts attempted to revive BeOS as an open source project but it never really got off the ground. The last remaining sign of life is Haiku, "a fast, efficient, easy to use and lean open source operating system inspired by the BeOS that specifically targets personal computing." There is little indication that anyone uses it, but new releases continue to appear.

Meanwhile as BeOS was dying, inferior operating systems such as Windows (and in the 1990s, Windows was really a pretty inferior product), prevailed. The better operating system disappeared.

It was not inevitable that BeOS would fail. One can easily imagine a scenario where Apple embraced it and it thrived. The reason for its failure was not intrinsic to the software itself, which was excellent, but to business decisions made by the Be Inc and Apple managements.

A similar tale can be told about WebOS – an attempt by Palm, the company that essentially created handheld computers (or at the least the first commercially successful ones) to get into the smartphone market.

Palm announced WebOS in early 2009 to great fanfare and it was immediately embraced by a community of enthusiasts. Among its designers were a number of key former Apple employees and the system had a number of advantages when compared to Apple's smartphones as well as the then-popular alternatives (Blackberry and Symbian). For example, it allowed genuine multi-tasking.

In June 2009, the first phones using WebOS – the Palm Pre followed by the Palm Pixi – were released by

Sprint in the United States. But sales were poor and Palm quickly sold WebOS off to the much larger and more established computer manufacturer Hewlett Packard (HP). It must have seemed like a good idea at the time.

HP attempted to revive WebOS as an operating system for tablets, but rapidly abandoned this after poor sales for its Touchpad device.

Eventually HP sold WebOS to the Korean firm LG Electronics where it is reportedly dying a slow death.

There are lessons here for the people behind Firefox OS, and also for those of who may choose to use it and build apps for it.

There are a number of parallels between Firefox OS and BeOS and WebOS which should set alarm bells ringing.

Even if Firefox OS is a really great idea (as was WebOS just a few years ago) and even if it turns out to be a better product than its competitors (as BeOS probably was), that doesn't guarantee its success.

Furthermore, the fact that we need an open source alternative to the Apple/Google duopoly doesn't necessarily mean that Firefox OS is the one that will survive. There are already several alternatives emerging.

Open source alternatives to Firefox OS

The world of mobile phone operating systems is dominated by the big two, with Microsoft's Windows Phone currently a poor also-ran with a market share below 10%. But below the surface there is a ferment of activity, with lots of alternatives under development. It is a fair bet that most of these will lose out. But it's worth remembering that Android was a tiny Californian start-up ten years ago, and that eight years ago it was a broke tiny start-up when Google bought it. Also, the very existence of so many projects suggests that lots of smart people think that there is something wrong with the status quo, and that there is an opportunity to disrupt it.

For this book we've identified four main alternative currents (apart from Firefox OS, which we are focusing on).

Sailfish

Developed and promoted by Finnish company Jolla, Sailfish is a lineal descendant of the MeeGo operating system once promoted by handset maker Nokia and chipmaker Intel. MeeGo, hosted by the non-profit Linux Foundation, represented an attempt by these two companies to build on their separate Linux-based handset operating system projects, and to head off the momentum behind Google's efforts. Intel was also concerned about Microsoft's lack of support for its own Atom processors.

But some of the main backers behind MeeGo jumped ship. In 2011 Intel decided to collaborate with Samsung on Tizen instead (see below) and Nokia has now sold its phone-making division (the bulk of its business) to Microsoft.

Its eventual failure proves that even the involvement of corporate powerhouses (and you don't get any bigger than Nokia and Intel) doesn't guarantee success. Indeed the fate of Nokia, once the dominant company in the global mobile phone business and now making its exit, is an eloquent demonstration of just how powerful is the Apple-Google duopoly.

However, MeeGo had by now attracted a community of developers, and these decided to soldier on with a "fork" of the MeeGo operating system which they called Mer.

A group of former Nokia employees formed Jolla as a new company to continue the development of Mer, which now forms the main basis for Sailfish.

This is all beginning to sound very Biblical, with MeeGo begat Mer and then Mer begat Sailfish, and maybe by the time you read this book, Sailfish will have a new name. That's just how these things work.

The advantages of Sailfish include:

- It's much closer to true Open Source than Android – not something that matters to the average user, but important for some independent developers. Sailfish phones will also run Android applications.
- It's based – like Firefox OS – on the HTML5 standard, again making application develoment easier (and there is a bunch of tools and testing software to help with this) and increasing the functionality of the browser and the possibility of web apps.
- Jolla plans to be a manufacturer too – so there won't be issues with getting the OS onto phones – and it claims that it can draw on the years of experience that Nokia has in optimizing software to operate on mobile devices
- Access for the US government – and perhaps other governments too – may be a little bit trickier because Jolla won't have any servers in the US

Tizen

When Intel stopped supporting Meego it transferred its loyalties to Tizen.

Intel wants to secure its own future as a chipmaker in a rapidly-changing world. Whereas it could once rely on its relationship with Microsoft to ensure that its chips would always be in demand for PCs, the new world is increasingly dominated by other kinds of devices (like tablets and smartphones) and these are often based on other kinds of chip – notably those produced to designs created by the British company ARM Holdings.

Tizen's other main backer is Korean electronics giant Samsung, now the main manufacturer of Android handsets. Samsung is doing very well out of Android, but it seems to want to keep its options open. After all, Google might one day change the terms on which it licenses Android – or Samsung might decide that Android does not offer it enough opportunities to create a different experience for its customers. There was an earlier in-house attempt to create an OS (called Bada) which has now been merged into Tizen.

Like Sailfish, Tizen is hosted by the Linux Foundation, and like other attempts to create momentum behind an alternative OS, there is a community of prestigious backers – this time including telecoms equipment manufacturers Huawei and NEC, Panasonic, and network operators Orange Telefonica and Vodafone.

But there have been persistent rumours that it is not really going anywhere, and there is not much to show for it so far. Dates for releases of devices keep being pushed back.

For those who care about such things, Tizen, like Android, is not fully Open Source, even though it is based on Linux and on many open source components. The licensing model has been described as complex, and the software developers kit is published under a Samsung licence rather than an Open Source one.

Ubuntu

Created by a UK-based software company Canonical, whose founder is South African entrepreneur Mark Shuttleworth, Ubuntu is the most successful version of Linux desktop software.

To be scrupulously fair, that's not saying much. Even Linux boosters admit that all versions taken together have less than a 2% share of the desktop operating system market. And Linux purists don't like Ubuntu all that much, because the company is sometimes seen as not playing by the rules of the overall Linux community.

On the other hand, Ubuntu has an established following of millions of users, a community of developers and forums, and a good track record of working with a wide range of PC hardware.

In January 2013 Canonical announced that it had been working on Ubuntu Touch, a version for use on smartphones and tablets, and a developer version was demonstrated and released at Mobile World Congress, the mobile industry's annual Barcelona trade show, in February. It is of course Linux-based, and it uses some components developed for the other alternative operating systems. The early versions certainly look attractive, and it won some industry awards, including tech media website CNET's "best in show" award for the Barcelona event.

Naturally, Canonical has set up a carrier advisory group to line up mobile network operators behind its efforts, and it has captured some big names including US operators Verizon and T-Mobile USA, Telecom Italia, Korea Telecom and China Unicom.

Telefonica, the ringleader of the carrier group behind Firefox OS, has also joined – showing both that everyone in the industry is seeking to keep their options open and that the act of joining one of these advisory groups is not proof of a high-level commitment. So the participation of some handset manufacturers, including LG from Korea, should not be taken as conclusive proof that devices are already in the works.

Perhaps because of this, Canonical concieved a rather daring plan to crowd-fund its own handset develoment. It sought to raise around $32m to build 40,000 of its own handsets as proof of concept; in the event it managed to secure pledges worth almost $13m, a record for the Indiegogo site but still short of its target. It has since said that it will nevertheless release the software in October 2013, so that enthusiasts can install it on to compatible devices. It will stay that way unless Canonical finds a manufacturing partner.

The proposed Ubuntu phone, dubbed a "super-phone" was intended to be perhaps the most high-spec device ever made, with a multi-core processor, sapphire crystal display, two cameras, 8GB of RAM and 128GB internal storage.

That figure of 128 GB is double the capacity of the most expensive version of Apple's latest iPhone.

The first, cheap Firefox OS phones offered far smaller capacity, but unlike Apple's devices, allowed the addition of an SD card of up to 32 GB.

Canonical was placing its bets on the improved

hardware available to make mobile devices, and arguing that the current generation of smartphones were punching below their potential weight.

It also argued that the super-phones would be capable of replacing desktop PCs – when complemented with an appropriate docking station that could connect to power, full size monitors and keyboards – and that this made them suitable for use in a business environment.

The philosophy behind the super-phone represents a departure from Canonical's comfort zone. If the company was seeking to draw on the loyalty of its existing customer base of Ubuntu enthusiasts, this was a funny way of showing it. One of the strengths of the desktop OS is its ability to make the most of old and under-powered hardware. The community of users is strongest in emerging markets, notably India and China. There might be a market for super-phones in places like this, but there might well not be.

Still, at least the super-phone vision is distinct, and offers the possibility that Ubuntu Touch might be able to find a separate niche alongside both the big two and other alternative operating systems.

Looking at Sailfish, Tizen and Ubuntu, one might wonder which – if any – of these will survive, and produce actual phones.

One possibility is that none of them will, but another is that some or all might.

The world of Linux distributions (known as "distros" to enthusiasts) is not limited to Ubuntu, and quite a few compete with it.

So one can imagine a future with not just the current duopoly of Apple and Google, but several competing open source systems, with Firefox OS being just one of them.

Android Mods

Finally, if you want an alternative to Android, how about . . . Android?

There are lots of discussions on the internet about whether Android is really open software or "free" (as in freedom, not as in free beer) software.

It's not a straightforward question, and to anyone who is not a particular kind of software geek the argument seems a little like those theological disputes about angels and the points of a pin.

Android is based on Linux, which is free and open source, but it's not based on GNU, which is where the free software movement seem to place most of their hopes. Android is free, in that Google doesn't charge manufacturers a licence fee, but it doesn't use the licensing framework favoured by free software partisans and it doesn't make all of the source code available.

The possibility that Google might one day start to charge for access to the Android OS is said to be one of the reasons why some of the biggest manufacturers maintain a favourable disposition towards the non-corporate beardy-weirdy alternative OS. And the hardware drivers that Android uses to enable it to talk to components in the phone and to the mobile network are not free or open – they are often proprietary. Nor is the Google Play Store, where Android users get their apps, free and open.

The philosophical aspects of this, and the details of where different flavours of open-ness fit together and overlap, are beyond the scope of this book.

Why should social change or union activists care? After all, most normal people just want their stuff to work and don't think about what goes on under the hood. That's why iPhones are so popular.

Well, Richard Stallman, a founder and guru of the Free Software Foundation, puts it well:

"On most Android phones, this [proprietary] firmware

has so much control that it could turn the product into a listening device. On some, it controls the microphone. On some, it can take full control of the main computer, through shared memory, and can thus override or replace whatever free software you have installed. With some, perhaps all, models it is possible to exercise remote control of this firmware to overwrite the rest of the software in the phone. The point of free software is that we have control of our software and our computing; a system with a back door doesn't qualify. While any computing system might *have* bugs, these devices can *be* bugs."

And if this sounds like a paranoid fantasy, even after all of Edward Snowden's revelations, then Stallman goes on to relate an anecdote from former British ambassador Craig Murray, about an intelligence operation that remotely converted an unsuspecting target's non-Android portable phone into a listening device.

Of course, other Android applications already spy on you in a slightly less invasive way, as you agreed when you installed them. Some network operators use the capabilities of the Android system to stop you from installing applications that they don't like, such as the ability to use your phone as a modem or WiFi hotspot without their permission.

But where there is a geek there is a way.

There are multiple projects to open up Android, to modify it so that it would be much more free and open. The purest in spirit of these Android mods is probably Replicant (named after the androids in the film Blade Runner), which is intended to be fully "free", is supported by the Free Software Foundation and developed by a community of coder enthusiasts. Replicant ships with the F-droid client as an alternative to the Google Play store, so that users can download and install entirely free software applications. It works on some devices but not others, and there is an updated list on the Replicant website. Installing is a non-trivial exercise, though it won't deter a really determined geek. But if you are that geek, then you've already passed beyond the scope of this book.

Another alternative version of Android has the catchy name of Cyanogenmod, also intended to be free and open source, and with the promise of some more obvious user benefits. Cyanogenmod is based on the latest release from Google but re-engineered to work on legacy devices and improved through the work of multiple developers. It points out that some features developed by its community eventually find their way into the formal Google release – though some others, particularly those which maintain the users' control over their own devices, may not. Cyanogenmod claims millions of installations, and says it is the most popular Android "aftermarket" OS.

In some cases Cyanogenmod is reputed to make your phone work better by pushing its underlying hardware harder, and to offer some features and functions – like a new kind of "launcher" - that's not in the mainstream version of Android. There is a larger set of supported devices, and installing, though still not trivial, is said to be somewhat easier.

Unlike Replicant, though, Cyanogenmod is no longer a loose community of developers but a fully-fledged West Coast start-up company, with venture capital backing. It has now entered into a manufacturing partnership with Chinese handset maker Oppo and has demonstrated its first phone with the OS pre-installed. This is not the same as having a big-name backer like Samsung or ZTE, though some of these are quietly supportive and have made it easier to install an aftermarket OS.

Conclusion

OK, we know we lost you there a bit with Cyanogenmod and no, we can't pronounce it either.

So let's review the main points and try to do this in the least geeky way we can.

First of all, there is a problem with the current situation with smartphones and tablets – they're expensive, the creation of apps for them is expensive and difficult, there are concerns about censorship and privacy, and so on.

The phones and tablets we use are made in factories where workers enjoy few rights and are disasters from an environmental point of view.

A number of interesting initiatives have emerged in the last couple of years, not least of them Fairphone and Firefox OS, and we've tried to honestly discuss the strengths and weaknesses of them.

Knowing that it's nearly impossible to predict what technologies will prevail and which will disappear (if we could do that, we'd be a lot richer than we are today), we've tried to survey the alternatives to Firefox OS which have emerged in recent months – including

Sailfish, Tizen, Ubuntu and the various Adroid derivatives.

We think that one or more of these might well survive and offer a cost-effective, flexible and open alternative to the proprietary systems Google and Apple offer us today.

As campaigners for a better and fairer world, we're always on the lookout for new tools, new technologies, that will strengthen our trade unions and campaigning organizations.

Two decades ago, email and the web were new, untried and untested, and yet we advocated their adoption and use by unions, because we saw the potential.

The same is true today with mobile technology.

Smartphones and tablets are increasingly inexpensive, easy to use, and in the hands of ordinary people.

That's why we need to show an interest in them, to develop applications that run on them, and maximize our use of them as tools for social change.

Firefox OS is not a sure thing. But if it does succeed, offering a low-cost, open alternative to the existing Apple/Google duopoly, it will be a very good thing.

What now?

If you're a union member, or a member of a campaigning organization, you should press your colleagues to **plan on designing web apps that will make use of new operating systems like Firefox OS**. You can explain that apps like these will be cheaper and easier to make than the apps currently running on iOS and Android. And that our organizations can make use of the fact that we have among us many, many web developers who don't need to learn new skills to create apps that will work on these phones and tablets.

You should also be asking your union or organization to **ensure that its existing website is responsive** – meaning that it will render perfectly on a small screen and not only on a desktop PC.

When your organization is thinking about **buying new phones or tablets** for staff or officers, they need to keep in mind that the phones we have been talking about in this book are considerably cheaper than some of the existing Apple and Android products. Even the "ethical" FairPhone is about 40% of the price of a high-end iPhone and your union or organization can have about a dozen of the entry-level Firefox OS phones for the price of one sleek new Apple phone.

If you're a techie, you should be keeping your eye on these developments, and learning how to transform existing websites into web apps that will run on these new devices.

Just a few years ago, if someone had suggested that Blackberry, which dominated the world of smartphones, was going to effectively disappear, they would have been thought crazy. But that was the effect of the launch of Apple's phone just a few years ago.

After that, anyone who suggested that the mighty Apple might be toppled by Google's Android system would also have been thought out of touch. But that too seems to have happened.

No one can predict what will happen next, but for those of us serious about using the new communications technologies to promote social change, we need to be careful not to miss what's coming next.

Acknowledgments

We'd like to thank Sean Ansett of Fairphone, David Wood of the Ubuntu Phone carrier group, and the Mozilla Foundation.

www.ingramcontent.com/pod-product-compliance
Lightning Source LLC
Chambersburg PA
CBHW071618170526
45166CB00003B/1105